# Alien Life and Dark Plasma:
## What Makes You Alive and Self Aware?

## Igor Kryan

ISBN 978-0-359-13417-5

www.youtube.com/c/IgorKryan

Printed in the United States of America

## Intro

What does it mean to be alive? If you ask a chemist, the answer will be - to contain DNA or RNA, or to be a protein. If you ask a psychologist, the answer will be - to be self-aware. If you ask a biologist, the answer will be - the capacity for growth, reproduction, functional activity, and continual change preceding death. If you ask a philosopher, the answer will be - there is no exact definition of what it means to be alive.

But really, what makes you alive? Complex organic chemistry reactions inside you body. But the fungus inside an old log also alive by this definition. So what makes you different from fungus? Self awareness. And the self awareness arising from your brain neuron cells connections according to all text books. What if I told that the only thing your self awareness has to do with your neurons in your brain is nothing more than a power generator has to do with high voltage wires. That's right, your brain is just a bunch of wires powered by a generator far away. Brain - no mater how complex it is - just wires that enable your body to move, but self awareness and true life including alien life originates from a very different source.

# Content

## Chapter 1. Alien Crystals.

For over 70,000 years, strange life forms have been living inside gigantic crystals at Mexican caves. A few years ago NASA researchers finally managed to revive some of them, leading to a astounding speculation on what these extremophile, "alien" organisms could teach us.

For many years, a team led by director of NASA Astrobiology Institute Penelope Boston, has explored Naica Mine in Chihuahua, Mexico searching for extremophiles, organisms that thrive in the most intense and strange of environments when it comes to extremes: ranging from acidity to temperature, to lack of nutrients, to lack of sunlight.

Caves as massive as underground cities can be found at this location in Mexico. However, the most interesting part about the mines and caves is definitely the fact that they are filled with gigantic gypsum crystals that make you feel like you're the size of a ant.

Not only are those giant gypsum crystals visually striking to the point of being gigantic, but tiny "bugs" were discovered inside them, in a preserved state of "geolatency," where living organisms stay stuck in geological materials for extremely long time.

"Much to my surprise, we got things to grow," said Dr. Boston. "It was laborious. We lost some of them – that's just the game. They've got needs we can't fulfill. That part of it was really like zoo keeping."

While they were described as "bugs," mostly bacteria were found, and it was over 100 different kinds of them. They were trapped straight inside the crystals, and they are thought to have been trapped there for between 12,000 and 70,000 years. Over 90% of these life forms have never been seen before at all.

The beautiful cave system lies directly above a quite large pocket of volcanic magma. Geothermal heating raises the temperature in there as high as 150 degrees Fahrenheit, or 70 degrees Celsius. Some astrobiologists call it "hell."

The heat allowed for those giant crystals to form. According to Dr. Boston:

"Most life could not survive there but scientists have discovered some organisms have evolved to feed on the sulphides, iron, manganese or copper oxide in the cave."

"They're really showing us what our kind of life can do in terms of manipulating materials," continued Dr. Boston.

"These guys are living in an environment where there's not organic food as we understand it. They're an example at very high temperatures of organisms making their living essentially by munching down inorganic minerals and compounds. This is maybe the deep history of our life here.

It's this type of thing that increase the probability of alien life existing on other planets and in other solar systems. Like on our planet Earth, just about every conceivable type of life form that can exist, does exist. Think of a shape, or a feature of some creature: it exists. Somewhere in the ocean, it exists. Somewhere out there in the savanna or in the mountains, that life form thrives if you can imagine it."

## Chapter 2. Remember Your Past Life?

You might rightfully state that cave crystal life is a biological life but not an intelligent life like you and me. But what really makes you different from that cave bug? You might say it's you life experiences and everything you learned over the course of your life.

However, in 2012 I published book called "Creator's Riddle: Darwin vs. God" where I described a couple of simple experiments: Take a newborn calf of stock that used to cattle grids but who never seen one and introduce it to lines painted on the road resembling grid. Calf will not cross them. Or take a newborn chick and place it in room with a hawk. It frantically tries to find cover but when it meets another non predatory bird for the first time - the same chick is completely comfortable. How has this knowledge been communicated?

It seems that you are not everything you learned over the course of your life but something much more. My daughter Alisa was only a couple days old, when some relatives arrived to see a newborn baby, coming one by one to the cradle smiling at her. Alisa was smiling back at them, until my dog Nikita decided to check what all that commotion is all about. He put his hairy face into her cradle. Alisa's good mood changed instantly. She was trying to hide and was crying in fear. Nikita was a very adorable small Lhasa Apso but from my newborn daughter's perspective he was a terrifying monster.

The incident made me seriously think how two day old baby, who knows nothing and can barely see, can also tell the deference between humans, toys and animals in a fraction of the second. And there was only one explanation: our identity leaves a permanent mark on genome. We pass onto our descendants much more than eye color. It has been already estimated that over 50 percent of known personality traits are preprogrammed before our birth.

You might say that what we experience as newborns is part of our genetic memory - a memory present at birth that suppose to exist in the absence of real life experience, and is incorporated into the genome over long spans of life. It is based on the idea that common experiences of species become programmed into genetic code.

However, there is one big flaw with this model. For those preprogrammed memories to exist, brain connections need to be permanent and stable and the brain is not. Nearly all brain's molecules, including those that form the neural connections thought to be involved in memory, are replaced every few months. It is physically and chemically impossible that long lasting memories that transcend though the generations of species are stored on such impermanent medium.

Furthermore, lower species, for example, butterflies have more photo-receptors in their visual system than we do, making their visual systems way better than our eyes in certain respects. However, while we have a big occipital (visual) lobe in our brains to process visual information the butterfly does a superior job with brains that are mere specks - less then two millimeters in total size. Scientists have no explanation as to how they do it. Plants and microorganisms likes amoeba go about their complex activities without having any brains at all. The brainless single-celled slime mould solves mazes every time it is tested. Where is this intelligence and information processing coming from? Mother Earth? The life-forms lower down the evolutionary ladder (with no or tiny brains) may often be using non-local intelligence to compensate for deficits in their biological cognitive system. Ants and bees would make the best examples.

Autistic savants, whose brains are impaired, also use this non-local intelligence to compensate for their deficits. Neuroscientists believe that autistic savants have access to regions in the brain that function like supercomputers. But where is this supercomputer located in our brain?

Across the world, there are hundreds of unfortunate cases of people with hydrocephalus - a condition where large cavities form in the brain. There are also hundreds of cases where people have either been born with an underdeveloped brain, or have had large areas of their brain damaged in an accident or removed surgically. Nonetheless, many of them function normally. Some scientists believe that the remaining cortex takes over the functions. However, this explanation becomes questionable especially in cases where hardly any brain is left in the skull.

Some scientists go even farther and suggest that these memory molecules might store information themselves, that each individual neuron contains memory. However, the usual scientific concept of our genetic code is something fixed at the beginning of our lives, not something that gets re-written all over again on a daily basis. Almost no cell in your body is designed to live over 10 years, yet humans live well into their 70s, 80s and even 100s thanks to the fact that our sell can reproduce themselves (some daily like epithelium, some once in 10 years like brain neurons). From biological and physical perspective, you were a completely different person assembled from different cells 10 years ago than you are today. The brain neurons are the only cells which were not fully replaced with new ones for the last 10 years allowing you to store lifetime experiences and memories. Yet, they also die and regenerate (as it was recently proven). This process also causing you to forget some distant memories.

To explain this phenomena we need to look not inside the non-existent brain but outside. We need to look at the super-cortical structures which interface with Earth's noosphere -global information filed. Carl Jung popularized the idea of a collective unconsciousness that we all are plugged into, and suggested it as the repository of memories and universal archetypes. Carl Jung's theory points to the presence of preprogrammed memories un a collective unconscious. If so, how do those memories travel to and from our brains? Keep reading and you shall find out.

## Chapter 3. Ball Lighting.

Believe it or not, but until 2014 most scientists denied that Ball Lighting phenomenon actually exist, despite numerous accounts from around the world. The first optical spectrum of a ball-lightning event, published in January 2014, included a video at high frame-rate.

One early account was reported during the Thunderstorm near church in Devon, in England, on 21 October 1638 following happened: "Four people died and approximately 60 were injured when, during a severe storm, an 9-foot (3 m) ball of fire was described as striking and entering the church, nearly destroying it. Large stones from the church walls were hurled into the ground and through large wooden beams. The ball of fire allegedly smashed the pews and many windows, and filled the church with a foul sulphurous odor and dark, thick smoke.

The ball of fire reportedly divided into two segments, one exiting through a window by smashing it open, the other going though the walls of the church. The explanation at the time, because of the fire and sulphur smell, was that the ball of fire was "the devil" or the "flames of hell". Later, some blamed the entire incident on two people who had been playing cards in the pew during the sermon, thereby incurring God's wrath."

November 4, 1749 Admiral Chambers on board the Montague ship wrote: "I was taking an observation just before noon...I observed a large ball of blue fire about three miles distant from them. They immediately lowered their topsails, but it came up so fast upon them, that, before they could raise the main tack, they observed the ball rise almost perpendicularly, and not above forty or fifty yards from the main chains when it went off with an explosion, as great as if a hundred cannons had been discharged at the same time, leaving behind it a strong sulphurous smell. By this explosion the main top-mast was shattered into pieces and the main mast went down to the keel. Five men were knocked down. Just before the explosion, the ball seemed to be the size of a large mill-stone."

An English journal reported that during an 1809 storm, three "balls of fire" appeared and "attacked" the British ship HMS Warren Hastings. The crew watched one ball descend, killing a man on deck and setting the main mast on fire. A crewman went out to retrieve the fallen body and was struck by a second ball, which knocked him back and left him with mild burns. A third man was killed by contact with the third ball. Crew members reported a persistent, sickening sulphur smell afterward."

On April 30, 1877, a ball of lightning entered the Golden Temple in India, and exited through a side door. Several people observed the ball, and the incident is inscribed on the front wall of Darshani Deodhi.

On November 22, 1894, an unusually prolonged instance of natural ball lightning occurred in Golden, Colorado, which suggests it could be artificially induced from the atmosphere. The Golden Globe newspaper reported, "A beautiful yet strange phenomenon was seen in this city on last Monday night. The wind was high and the air seemed to be full of electricity. In front of, above and around the new Hall of Engineering of the School of Miners, balls of fire played tag for half an hour, to the wonder and amazement of all who saw the display.

On May 22, 1901 in the Kazakh city of Uralsk in the Russian Empire (now Oral, Kazakhstan), "a dazzlingly brilliant ball of fire" descended gradually from the sky during a thunderstorm, then entered into a house where 21 people had taken refuge, "wreaked havoc with the apartment, broke through the wall into a stove in the adjoining room, smashed the stove-pipe, and carried it off with such violence that it was dashed against the opposite wall, and went out through the broken window".

On April 29, 1925 in Germany multiple witnesses saw a silent ball move along a telephone wire to a school, knock back a teacher using a telephone, and bore perfectly round coin-sized holes through a glass pane. Over 700 feet (250 meters) of wire was melted, several telephone poles were damaged, an underground cable was broken, and several workmen were thrown to the ground but unhurt.

Hundreds of well trained military Pilots in World War II described an unusual phenomenon for which ball lightning has been suggested as an explanation. The pilots saw small balls of light moving in strange trajectories, which came to be referred to as foo fighters. According to almost all pilots who saw them - the balls seemed to be following the planes and exhibiting some sort of either intelligence or at least formations and strategy.

Submariners and NAVY officers in the Second World War gave the most frequent and consistent accounts of small ball lightning in the confined submarine atmosphere. There are multiple accounts of repeated inadvertent production of floating explosive balls. Official NAVY explanation was: "When the battery banks were switched in or out, especially if mis-switched or when the highly inductive electrical motors were misconnected or disconnected - the balls are produced." However, attempts later to duplicate those balls with a surplus submarine battery resulted in several failures and an explosion.

And this is the more recent statements from Airline Pilots and passengers to FAA "We seated near the front of the passenger cabin of an all-metal airliner (Eastern Airlines Flight EA 539) on a late night flight from New York to Washington. The aircraft encountered an electrical storm during which it was enveloped in a sudden bright and loud electrical discharge (00:05 h EST, March 19, 1963). Some seconds after this a glowing sphere a little more than 20 cm in diameter emerged from the pilot's cabin and passed down the aisle of the aircraft approximately 50 cm from me, maintaining the same height and course for the whole distance over which it could be observed."

Finally, December 15, 2014, flight BE-6780 (Saab 2000) in the UK experienced ball lightning in the forward cabin exiting the aircraft and producing lightning strucking the aircraft nose.

In January 2014, scientists from Northwest University in Lanzhou, China published the results of recordings made in July 2012 of the optical spectrum of what was thought to be natural ball lightning made by chance during the study of ordinary cloud–ground lightning in Tibet. "At a distance of 900 m (3,000 ft), a total of about 3 second of digital video of the ball lightning and its spectrum was made, from the formation of the ball lightning after the ordinary lightning struck the ground, up to the optical decay of the phenomenon. Additional video was recorded by a high-speed (3,000 frames/sec) camera, which captured only the last 1 second of the event, due to its limited recording capacity. Both cameras were equipped with spectrographs. The researchers detected emission lines of atomic plasma and also containing silicon, carbon, calcium, iron, nitrogen and oxygen — in contrast with mainly ionized nitrogen emission lines in the spectrum of the regular lightning. The ball lightning traveled horizontally across the video frame at an average speed equivalent of 8.6 m/s (28 ft/s). It had a diameter of 5 m (16 ft) and covered a distance of about 15 m (49 ft) within those 1.64 s.

From the spectrum, the temperature of the ball lightning was assessed as being lower than the temperature of the parent lightning

(<15,000–30,000 K (14,700–29,700 °C; 26,500–53,500 °F)). The observed data are consistent with vaporization of soil as well as with ball lightning's sensitivity to eclectic fields, therefore, being a plasma balls."

Stable considerable size and long lasting Lighting balls were produced in laboratory only once by Nicola Tesla. He could artificially produce over 1.5-inch (40 mm) balls and conducted some demonstrations of his ability, but he was truly interested in higher voltages and powers, and remote transmission of power, so the balls he made were just a curiosity for him.

When we were children we decided to reproduce Nicola Tesla experiments and capture "a living thunder ball." During huge thunderstorm we constructed 30 feet (10 meter) long lighting rod attached to the large 3 feet (1 meter) sphere made with mesh wire and placed on top of the building under construction on the high hill in the city of Kiev. Knowing that ball lightning's attracted to electric fields, we also connected 380 Volts 15 Amp power cable from the transformer to our tower and waited nearby in the same building for lighting to strike. When it finally happened, the lighting destroyed the entire tower along with the transformer making the wires glow for a few seconds but no lighting ball was produced to our disappointment. Later on I realized, that had we succeeded to produce 1 meter or 3 feet thunder ball, I wound not probably be around to write this book anymore.

But the fact is - Lighting balls do exist and they composed of gas plasma and also containing silicon, carbon, calcium, iron, nitrogen and oxygen - basic building blocks of life.

## Chapter 4. Plasma Crystals.

There are many theories how lighting balls are produced. Ranging from Black Hole Hypothesis stating that ball lightning is the passage of microscopic primordial black holes through the Earth's atmosphere, to Transcranial magnetic stimulation theory.

Cooray and Cooray (2008) stated that "the features of hallucinations experienced by patients having epileptic seizures in the occipital lobe are similar to the observed features of ball lightning. The study also showed that the rapidly changing magnetic field of a close lightning flash is strong enough to excite the neurons in the brain. This strengthens the possibility of lightning-induced seizure in brain of a person close to a lightning strike, establishing the connection between epileptic hallucination mimicking ball lightning and thunderstorms."

More recent research with Transcranial magnetic stimulation theory has been shown to give the same hallucination results in the laboratory and these conditions have been shown to occur in nature near lightning strikes. This hypothesis fails to explain observed physical damage caused by ball lightning or simultaneous observation by multiple witnesses.

Theoretical calculations from University of Innsbruck researchers suggest that the magnetic fields involved in certain types of lightning strikes could potentially induce visual hallucinations resembling ball lightning. Such fields, which are found within close distances to a point in which multiple lightning strikes have occurred over a few seconds, can directly cause the neurons in the visual cortex to fire, resulting in  magnetically induced visual hallucinations."

In other words, one theory out of about 100, telling that Lighting balls are product of our brain. What if I told you contrary, that our brain is the product of Lighting balls. Sounds insane? Not so fast.

All other 100 theories agree that Lighting balls are plasma. And "Plasma is the fourth state of matter. When gas is superheated, electrons are torn from atoms and become free floating. The gas then becomes ionized, carrying positive charge. This superheated mix of ionized gas and free-floating electrons makes up plasma. Stars are mostly plasma, as is about 99 percent of the matter of the universe, though plasma is much less common on Earth, where we're used to dealing with solids, liquids and gases. Besides being found in stars and in our Sun, plasma is transported by solar winds and magnetic fields, often coming into contact with dust clouds.

"When plasma comes into contact with a dust cloud, dust particles gather an electric charge by sucking up electrons from surrounding plasma. This core of electrons in turn pulls in positively charged ions, forming plasma crystals. In the simulations, which were performed on the International Space Station and in a zero-gravity environment at a German research facility, the plasma crystals sometimes developed into corkscrew shapes or even the double-helix shape of DNA. These helix-shaped crystals retain an electric charge demonstrating what the researchers called a life like self-organizing ability.

Once in helix form, the crystals can reproduce by diving into two identical helixes, displaying "memory marks" on their structures.

The diameter of the helixes varies throughout the structure and the arrangement of these various sections is replicated in other crystals, passing on what could be called a form of genetic code. They even seem to evolve. The formations become sturdier over time as weaker structures break down and disappear.

The universe is filled with massive clouds of dust and scientists have learned that this cosmic dust can, in the presence of plasma, create formations known as plasma crystals. In simulations involving cosmic dust, the researchers witnessed the formation of plasma crystals displaying some of the elementary characteristics of life - DNA-like structure, autonomous behavior, reproduction and even evolution."

If they exist in the simulated form, the researchers believe that the crystal organisms might be found in the rings of Uranus and Saturn, which are made up of small grains of ice. So they form DNA-like shapes, reproduce, pass on their structure or genetic code, "eat" plasma, evolve and die. But are these supposed organisms real life forms?

In July 2007, group of American scientists, in association with the National Research Council, issued a report recommending that scientists search for so-called weird life on other worlds, in space and even on Earth. Weird life is believed to be far different from life forms we're used to see. Weird life may be organisms that don't depend on water or that don't have DNA at all. Some people even believe that weird life existed on Earth in the ancient past and that it may still exist on this planet. In reality, scientists don't know what weird life is, but its presence has many reexamining notions of what alien life may be and where it might be found."

If the plasma crystals do exist in their simulated form, they live and develop at a pace at least a hundred thousand times slower than Earth's biological organisms. The question is then raised: given their fragility and slow pace of development, can they become intelligent or sentient?" This question will be answered in the next chapter.

## Chapter 5. Life and Reincarnation.

Three prominent scientists Tsytovich, Lozneanu and Sanduloviciu proposed that the plasma crystal cell was a precursor to the biological cell in the early Earth - acting as a template or mould for the biological cell to form. Now Igor Kryan is proposing that the lightning strikes frequent in early earth atmosphere attracted thunder balls or lighting balls containing silicon, carbon, calcium, iron, nitrogen and oxygen - basic building blocks of life and, therefore, sparked the organic biological life the plasma molds of cells.

Indeed, Plasma itself possesses Life-Like Qualities of Plasma: Bohm, a leading expert in twentieth century plasma physics, "observed in amazement that once electrons were in plasma, they stopped behaving like individuals and started behaving as if they were a part of a larger and interconnected whole. Although the individual movements of each electron appeared to be random, vast numbers of electrons were able to produce collective effects that were surprisingly well organized and appeared to behave like a life form. The plasma constantly regenerated itself and enclosed impurities in a wall in the same way that a biological organism, like the unicellular amoeba, might encase a foreign substance in a cyst. So amazed was Bohm by these life-like qualities that he later remarked that he frequently had the impression that the electron sea was "alive" and that plasma possessed some of the traits of living things."

The debate on the existence of plasma-based life forms has been going on for more than 20 years ever since some models showed that plasma can mimic the functions of a primitive cell. Minimal dark plasma cell-systems were generated within biosphere in the early Earth and were precursors for the existence of terrestrial life forms which evolved from these minimal plasma cell systems.

Plasma cosmologist, Donald Scott, notes that "...a [plasma] double layer can act much like a membrane that divides a biological cell". Plasma physicist Hannes Alfvén also noted the association of double layers with cellular structure, as had Irving Langmuir before him, who coined the term "plasma" after its resemblance to living blood cells.

David Brin's Sundiver also speculated on plasma life forms. This science fiction proposed a form of life existing within the plasma atmosphere of a star using complex self-sustaining magnetic fields. Similar types of plasmoid life have been proposed to exist in other places, such as planetary ionospheres or interstellar space.

These life forms would be as varied in scale, structure and intelligence as carbon-based life forms - as different as a microbe from a whale; a mosquito from a tiger; a giraffe from a crocodile; an ant from a human being. Their degrees of intelligence and awareness were as different as a centipede's awareness to the awareness and intelligence of humans. Some of these plasma life forms have interacted with us in the past either intentionally or unintentionally.

The entities that we have loosely identified as ghosts, angels, jinns, demons, deities, aliens, biological UFOs, fairies and sightings of the recently deceased, on the surface of the Earth, are characteristic of these predicted exotic plasma life forms from interpenetrating dark plasmaspheres or counterpart Earths. They constitute an ecology of plasma life forms that evolved throughout Earth's history and sometimes formed symbiotic relationships with the carbon-based life forms that we are more familiar with. Humans evolved carbon-based bodies that formed symbiotic relationships with some of these plasma life forms (indicating a type of symbio-genesis). When the carbon-based bodies died, the bioplasma bodies (resulting from the symbiosis) continue their existence and might recycle and fill up the other physical bodies. This is the only scientific theory that explain reincarnation and past life memories experienced not only by Buddhism an Hinduism devotees but by all religion followers as well as atheists.

## Chapter 6. Dark Matter.

A decade ago, astronomers have actually detected a 'universal web'. Vast filaments of hot gas tracing the web have been 'seen'. Astronomers using NASA's X-ray satellite observatory, Chandra, 'viewed' the filaments stretching for millions of light years through space, with one passing through our own galaxy. Astronomers say that the filamentary structures are so hot that it would generally be invisible to optical, infrared, and radio telescopes. These invisible filaments are detected only because higher density ordinary matter tends to accumulate and condense in these filaments - generating radiation which can be measured by scientists to confirm the existence of these filaments in intergalactic space. Being invisible, they are by definition components of 'dark matter and energy'. Dark matter and energy are invisible matter and energy that make up more than 99% of our universe.

What is Dark Matter? "Essentially, plasma of much higher energy particles (as predicted by supersymmetry theory) that emits electromagnetic waves beyond the known electromagnetic spectrum would be classified as invisible dark matter." The particles could even be charged but they would not be detectable by our current instruments. David Peat says, "It is indeed theoretically possible for a shadow universe to exist in parallel to our own. While we would feel its gravitational effects, this shadow universe would be otherwise invisible. Photons (light) from the shadow group would have no interaction with the matter in our universe."

Most of our current universe is in the form of plasma. Magnetic fields can be found in every region of space within our universe. Hence, we conclude that magnetic plasma is pervasive throughout this universe.

"Plasmas are not just the 'fourth state of matter' - they are really the first state in modern cosmology, and they continue to be, by far, the dominant state of visible matter in the universe; perhaps also of invisible matter as well if so-called 'dark matter' continues to remain unobserved and unexplained." - Timothy Eastman, President, Plasmas International.

Every galaxy, every solar system and every planet is essentially surrounded by this undetectable dark matter plasma. Carl Jung and Teilhard de Chardin views "the noosphere as the "collective consciousness" of human-beings which emerges from the interaction of human minds on Earth and is enriched as the population of human beings on Earth increases. According to Dark Plasma Theory the visible Earth sits inside a Jupiter-sized dark plasma sphere which is similar in size and shape to Earth's magnetosphere. The dark magnetic plasma generates filaments which synapse at nodes (with embedded vortexes) which are the analogues of the axons-dendrites and cell bodies in a human brain. These filaments in Earth's brain have also been identified as "ley lines" and the vortexes as "sacred sites" or portals in the general metaphysical literature. As in the cosmic brain, the Earth's electromagnetic plasma brain provides a suitable infrastructure for the rapid encoding of memories. Dark plasma, unlike ordinary plasma, is long-lived even at room temperatures.

The Akashic records that have been accessed by various readers of the human species are fairly recent memories encoded in Earth's brain. According to metaphysicist, Charles Leadbeater, when the observer is not focusing on them, the Akashic records simply form the background to whatever is going on – "reflecting the mental activity of a greater consciousness on a far higher plane which is accessible to us. Observing the dynamic and visual Akashic records would be like watching a larger brain's movie from a distance. We inhabit this much larger brain and are encoding memories not only within our own biochemical brain but within this dark noosphere."

This encoding process in Earth's brain may be taking place during REM (Rapid Eye Motion Sleep) via the left hemisphere of the brain and through the plasma body interface when our carbon-based body undergoes paralysis allowing our plasma body to decouple from it to communicate more freely with the dark neural networks that we are embedded in. This facilitates long term memory formation. Resonance between similar circuits in the human brain and Earth's brain may also play a part in reinforcing memories.

Most of the memories of the human species are encoded within a few kilometers of the Earth's crust - within what can be described as the "geocortex". These include autobiographical memories of individuals and various cognitive maps. In the wider space, memories of body maps of life forms, which play an important role in morphogenesis and the evolution of the species, are also be encoded. Associative memory ensures that thoughts relating to the same subject matter are accumulated – resulting in libraries, databases and specific types of environments – as recounted by readers of the Akashic records and ensures that "like attracts like". That forms the basis of Carl Jung's "collective unconscious" and the Buddhist "alaya" or storehouse of memories.

This dark plasma is similar to Wi-Fi network, continue to fill the brains of more primitive creatures we discussed before as well as humans with brain damage and autistic savants enabling them to perform and sometimes outperform ordinary humans in complex tasks.

## Chapter 7. Moon and Mars.

The same Dark Plasma phenomena is responsible for telekinesis. There is evidence that direct communication between physically separated human brains can and does occur. There are now seven published studies about this subject, six of which report significant findings. In one experiment, discussed by Dean Radin, light was flashed at the sender on one end. The receiver, who was inside MRI machine, showed activity in the occipital lobe (in the primary and secondary visual cortex) which correlated with the flashes of light.

These recent scientific studies provide persuasive evidence that there are deep invisible interconnections between human brains which are physically separated. This book proposes that Earth's brain has been the intermediary between these physically separated human brains.

Radin's analysis of Random Number Generator experiments showed strong evidence for attenuation due to distance between sender and receiver. He therefore believes that psi is not completely independent of distance. He cites Fiona Steinkamp's analysis of Extrasensory Perception ESP card guessing tests which also found a decline in effect sizes with increasing distance. The effect of distance on effect sizes cannot be accounted for by "quantum entanglement" models of telepathy which require immediate correlations and would be unaffected by distance. Furthermore, these modelsdo not allow actual information transfer.

It would be very interesting to see what will happen to human brains when they are torn apart from Earth Dark Plasma. The farthest earthling traveled is the Moon in 1971, which is still in the area partially covered by Earth magnetic field Earth Dark Plasma halo.

Yes, Moon landing happened not in 1969 but in 1971. It might sound something like from Yellow Press newspapers now called fake news but I guarantee that after reading will believe that knight in shining or better to say tarnished armor was found on the moon and I am not joking. The proudest moment in American and World History. For the first time man taking step on the ground outside of earth. But did it really happen? There are so many conspiracy theories about moon.

Indeed, why American flag is waving on wind while there is no atmosphere on the moon? Or this released by NASA original Apollo landing video with the big question who is filming it from outer space? Recently, NASA engineer came forward with a top secret video of the underwater facility saying that all space activities including most of the lunar landings were filmed there. Is this the ultimate evidence of great deception?

But don't jump to conclusions yet. And what about all astronauts who claim and even testified before congress saying there where on the moon? Did they committed perjury? Some of them are but they cannot be jailed for that. I remember the case in year 2000 when first President Vladimir Putin had a big dilemma what to do with the second Russian President Boris Yeltsin double when real Boris resigned. His double sincerely and genuinely believed that he is president of Russia and responded only to the name President Boris Yeltsin and he had no other identity in his head because of years of brainwashing and indoctrination by Russian FSB who prepared the perfect double for the first president. I do not know his fate - the mental ward is the optimistic version. The same story with all Apollo astronauts - put them on lie detectors and they will pass it stating they did walk on moon. But don't despair. After all, we did go to the moon. Not in 1969 but in 1971.

The first actual NASA attempt to land man on the moon was in 1970 Apollo 13 but mission was aborted following oxygen tank explosion, flew past the Moon without entering orbit and returned to Earth. And the first actual landing and return took place in 1971 with Apollo 14 and 15. I know it first hand from president Carter and now you know it too. United States was loosing space race to Soviet Union so we faked Apollo 11 and Apollo 12, failed Apollo 13 and only successfully landed and returned Apollo 14 and 15 in 1971 - two years after the first Lunar landing was announced.

However, we were not the first there. We were only the first to safely land and return humans in 1971. Soviet beat us there but they were not able to return from moon back to earth. September 14, 1968 one year before US first faked landing - Soviet were the first to deliver two turtles to the moon and then return - recovering them alive in Indian Ocean. I bet you did not know that - but check any US or Russian records and search for September 1968 Lavochkin Proton K/D mission Zond 6 - it's there. The difference between US and Soviet Union that Soviets would announce the mission to public only after it success. So the Soviets were the first to deliver a life form to the moon and then safely return it back.

July 13, 1969 - Russian Luna 15 with 1 cosmonaut hard landed on the moon. Radio contact was lost but it did land there. Panicking NASA realized that they are loosing Russians again, staged moon landing just a few days after. Staging was done badly and the original moon landing video tape was later officially destroyed by NASA - leaving only some copies with lots of inconsistencies in them. Russians were not able to contact their moon cosmonaut and after US made world wide celebration of the successful moon landing and return - Soviets decided to keep their dead cosmonaut in secret. They did notify NASA however, so later in 1970s Apollo astronauts mission objective was to find and bury Russian presence on the moon. They indeed found a medieval knight - with the only difference that his heavily armored knight costume was Soviet moon gear that indeed was looking like something from 14th century. The knight cosmonaut and the truth was buried on the moon. Now you know the story. Now you were vindicated: if you thought that moon landing was fake - yes, first two landings were staged. If you thought that you or your parents lived though the proudest moment in mankind history - yes, you or they did - we landed there in 1971.

However, the question of this chapter is not the exact date of moon landing but what exactly will happen to human brain when we land on Mars and loose not only earth protective electromagnetic field but dark plasma field as well? Will our wired receptors called brains connect to Mars much weaker alien dark plasma field and obtain the secret knowledge of planet's past? Or will they encounter living lightning balls - alien space plasma life-forms we wrote about earlier?

Many astronauts reported seeing those life forms in the space similar to foo fighters reported by Allied and German Air-force during World War II. One of them was original Mercury Astronaut Major Gordon Cooper and the last American to fly in space alone. On May 15, 1963 he was launched into space in a Mercury capsule for a 22 orbit journey around the world. During the final orbit, Major Gordon Cooper told the tracking station at Muchea (near Perth Australia) that he could see a glowing, greenish orb ahead of him quickly approaching his capsule. The UFO was real because it was picked up by Muchea's tracking radar. Cooper's sighting was reported by the National Broadcast Company, which was covering the flight step by step, but when Cooper landed, reporters were told that they would not be allowed to question him about UFO sighting.

Ten years earlier, in 1951 Major Cooper had sighted a UFO while piloting an F-86 Sabrejet over Western Germany. "They were saucer-shaped glowing discs at considerable altitude and could out-maneuver all American fighter planes. Major Cooper also testified before the United Nations: "I believe that these extra-terrestrials are visiting this planet from other planets... Most astronauts were reluctant to discuss UFOs. I did have occasion in 1951 to have two days of observation of many flights of them, of different sizes, flying in fighter formation, generally from east to west over Europe. For many years I have lived with a secret, in a secrecy imposed on all specialists in astronautics. I can now reveal that every day, in the USA, our radar instruments capture objects of form and composition unknown to us. And there are thousands of witness reports and a quantity of documents to prove this, but nobody wants to make them public. Why? Because authority is afraid that people may think of God knows what kind of horrible invaders. So the password still is: We have to avoid panic by all means."

"I was furthermore a witness to an extraordinary phenomenon, here on this planet Earth. It happened a few years ago in Florida. There I saw with my own eyes a defined area of ground being consumed by flames, with four indentions left by a glowing flying object which had descended in the middle of a field. Beings had left the craft (there were other traces to prove this). They seemed to have studied topography, they had collected soil samples and, eventually, they returned to where they had come from, disappearing at enormous speed... I happen to know that authority did just about everything to keep this incident from the press and TV, in fear of a panicky reaction from the public."

NASA astronaut Story Musgrave claims to have seen glowing eel-like tubes swim through space. In the interview above, he explains that he saw this "creature" on two separate occasions. While some immediately dismiss this as space junk, possibly some type of hose that detached from a spaceship, Musgrave is insisting that the white eel had its own propulsion technique.

Major General Vladimir Kovalyonok was part of a crew manning Salyut VI space station in 1981: "When I was working at the Salyut orbital station, I saw something strange in a porthole one day. The object was the size of a finger. I was surprised to see it was an orbiting object. It was hard to determine the size and the speed of an object in space. That is why I can not say exactly, which size it actually was. [My partner Viktor] Savinykh prepared to take a picture of it, but the UFO suddenly exploded. Only clouds of smoke were left. The object split into two interconnected pieces. It was reminiscent of a dumb-bell. I reported about it to the Mission Control immediately."

In 2005, Leroy Chiao was commander of the International Space Station. While on a spacewalk, Chiao saw white lights aligned in an upside-down check formation whiz right past him. Some people have posited that a string of fishing boats along the South American coast could explain what he saw, but Chiao was 230 miles above Earth when this happened. Those would have to be some impossibly strong boat spotlights to be seen from all the way up there.

In 2014, European Space Agency astronaut Samantha Cristoforetti was on her way to the International Space Station for the first time when she saw the normally gray ISS was bathed with glowing orange light. Cristoforetti was taken aback by the beauty of what she was seeing, and in a blog post, she wrote, "The enormous solar panels were inundated with a blaze of orange light, vivid, warm, almost alien." But none of the other astronauts saw this effect before.

John Glenn who flew on the Friendship 7 spacecraft in February of 1962, suddenly noticed something strange outside his window while in orbit. He immediately reported to NASA that he was watching what looked like a group of little glowing fireflies dancing outside his window. During the early Apollo missions, astronauts reported seeing "light flashes". The crews of later Apollo mission were warned about this and reported that they also saw strange bursts of light, flashes and glowing orbs.

Astronaut Alan Bean allegedly saw glass domes from a long-extinct alien civilization on the moon. In an interview, Bean described the space as looking like "black, patent-leather shoes" from the surface of the moon. Hoagland maintains, "Space should be velvet-black. It should be inky-black. It should be infinity, unending, deep, endless black. It shouldn't be shiny as I saw it."

The similarities between visions of trained Astronauts of 1950s and 2000s are astounding: shiny glowing flying objects of green, white or orange color, sometimes exploding. What are they actually seeing? UFOs from other planets or alien plasma bodies inhibiting space. Perhaps, the most of them saw nothing at all, but being partially torn away from earth dark plasma electromagnetic field, their brains started to pick alien dark plasma records of Mars and other planets making their brain to switch between our Earth reality and alien realities of the space.

The million dollar question still stands: what exactly will happen to human brain when we land on Mars and loose not only earth protective electromagnetic field but dark plasma field as well? Will our wired receptors called brains connect to Mars much weaker alien dark plasma field and obtain the secret knowledge of planet's past? Or will they encounter living lightning balls - alien space plasma life-forms that dozens of other astronauts saw?

Doctor Jay Alfred: "It is theorized that as brains evolve (in particular, the neocortex) and become larger (in proportion to the body) the more the life-form gets cut-off from non-local intelligence and has to rely on local intelligence in the brains housed within their bodies. Human beings, in particular, have become increasingly estranged from non-local intelligence. We create our own universes within our brains which compete with information coming from Earth's brain for our conscious attention.

However, evolution may have stepped in to correct this. The prefrontal and temporal lobes in the human brain underwent the most significant development during Human evolution. The right temporal lobe in the human brain could have evolved more significantly in Homo sapiens to circumvent complete blockage to Earth's brain to give our species an evolutionary advantage, specifically, unprecedented creativity in the animal kingdom.

However, the memories generated by our experiences are relayed into Earth's brain through the left hemisphere probably during REM sleep. On the other hand, we reconnect with Earth's brain when certain neural processes or circuits in our brains break down or is "switched off" intentionally (as in meditation, deep sleep, induced trances and the ingestion of psychoactive drugs) or unintentionally (as in reverie or brain damage due to accidents or medical conditions such as brain lesions, tumors and biochemical imbalances) while others are "switched on" in the right hemisphere (in particular the temporal lobe) to record the experiences and enable conscious recall. Hence, neural signals from the biological brain are output from the left hemisphere into the bioplasma interface which is then transmitted to Earth's dark plasma halo while input signals from Earth's brain arrive via the bioplasma interface to the right hemisphere."

In other words, question what exactly will happen to human brains on Mars, let alone in the distant galaxy, is not clearly answered yet. Igor Kryan best guess - it has its own earth like micro universe and will be able to survive by reconnecting it to alien dark plasma backed up by Universal global information field. But such astronauts will stop being purely humans and will become part aliens, thanks to the new information they brains will receive though the dark plasma electromagnetic field. Especially, since "Astronomers say they have discovered a giant magnetic and dark matter field that is coiled like a snake around a rod-shaped gas cloud in the constellation Orion." - Ker Than, Space 'Slinky' Confirms Theory with a Twist. The helical shape of the magnetic field around the gas cloud in the constellation Orion is believed to be caused by matter in the interstellar cloud moving in a straight line along the length of the filament. When this happens, it causes the magnetic field around the cloud to spiral around in a corkscrew pattern. When helical magnetic fields form in plasma, charged particles move along the field lines generating helical currents."

Should humans venture to the constellation of Orion one of two thing will happen - either exceptionally strong Orion plasma will implant new memories and new experiences on our brains, or wipe them clean, making our memories fuzzy, like most of the alien abductees report. There is a third possibility, however, Orion dark plasma field might be one living organism - either like Earth Gaya field on steroids or some much more complex life form that we cannot even begin to understand yet. But the fact is: ancient Egyptians definitely knew something about Orion that we only now begin to discover and they believed that their plasma souls travel to that Orion entity after bodily death.

## Chapter 8. Telekinesis.

The Global Consciousness Project uses Random Number Generator all over the world to record any apparently non-random movements during major news events. It was noted that prior to certain events, the Random Number Generator data would spike. The biggest spike was seen about two hours before the first 9/11 attack on the World Trade Center.

This provides some evidence of mass unconscious precognition indicating that human brains are tapping into a larger brain or neural network which processes time in a different and more expansive way. The psychics (who are really acting as mediums) often feel the pain of the victim in specific parts of their body and in some cases viewed the crime scene from the victim's or attacker's point of view. In other words, they were retrieving the autobiographical memories of the victims some unknown source. Observing a medium like Lisa Williams, who was proven correct multiple times, during a "reading" shows her not only recalling the deceased' intimate autobiographical memories but, most notably, also feeling the physical pain that the deceased suffered – whether it was due to a disease like cancer or a gun shot wound at specific sites in her body which correlates accurately with independent information. Many cases like this have been routinely reported and documented in forensic history.

These readings by mediums or psychics are similar in nature to readings of the Akashic records or the experience of undergoing a fast forward life review during a near-death experience. Persons undergoing a life review during a near death experience often feel the emotions of others who are affected by their actions (which would otherwise be unknown to them). For example, "a pilot during a war experienced the pain and anguish of a whole village when he unleashed his bombs on them, causing what we know today as Post Traumatic Stress Disorder. This can happen if the life review emanates from a higher brain centre which has access to the autobiographical memories of numerous other persons, many of whom may be unknown to the subject until the life review, but are linked in Earth's dark plasma electromagnetic brain through association pathways. Although wired neurons in the brain allows us to experience the emotions of others, the detailed nature of these experiences and the new information generated would suggest a genuine outside of the brain process, mediated by Earth's electromagnetic halo," states Doctor Jay Alfred.

In "group consciousness", telepathy and remote viewing, the Earth's "neural networks" are actively used by participants. According to Melvin Morse, "Remote viewing doesn't involve actually seeing something as much as it involves processing information through our right temporal lobe from the patterns of information contained in the universe. These patterns of information are the neuronal networks in dark plasma within which the visible Earth is embedded. This is one method that remote viewers may employ." Harman and Rheingold believe that the research on remote viewing suggests that "the creative/intuitive mind could be getting information in ways other than from the lifelong learning of the person". Many discoveries, scientific and artistic ideas come about when scientists or artists are not actively thinking about them for the same source as physics and mediums use.

In a similar way, when a person prays, signals are sent from his or her brain to different parts of the cortex in Earth's global information plasma field depending to whom the prayer is directed to. If it is directed to a well-known deity, the relevant associative memories, generated by human brains over the ages and found in fairly localized regions in the Earth's dark plasma brain with respect to that deity, are activated. Signals are then projected out through the pathways in Earth's brain and received by individuals who are also focusing on the same target and who may become participants of the intention of the prayer. Distant healing may also use such a process.

The question is how this Dark Plasma or Global Information field has formed? And the answer is "Minimal (dark) plasma cells developed in this 'dark biosphere' after Earth formed. Over more than 4 billion years these first cells evolved into complex dark plasma life forms. Plasma life forms they exhibit features commonly seen in ordinary standard plasma. Heraclitus was probably close to the mark when he hypothesised that "the soul was composed of a rare 'finer substance' which appeared like air or fire to him – two substances which, from a lay perspective, would bear a close resemblance to plasma. However, "these life forms are not composed of ordinary short lived plasma but non-standard plasma composed of dark matter particles, which is invisible to most humans just as dark matter is. Some of these life forms formed symbiotic relationships with members of the human species and co-evolved, integrating with the living carbon-based human body." In other words, we all are possessed by the global information field Dark Plasma.

## Chapter 9. Soul and Immortality.

Greek philosopher, Heraclitus, who lived in the sixth century BC, thought that the soul was composed of an unusually fine or rare kind of matter, such as air or fire and it does not disappear with death. On the death of the carbon based bodies, the dark bio-plasma bodies separate and return to their societies in the dark biosphere - Universe dark matter electromagnetic halo. It is these exotic life forms that are popularly called the 'souls' of human beings. It's not that far fetched to say that heaven and hell present in nearly all earth religions are real for the souls. Those symbiotic relationships of our brains with surrounding dark matter plasma life forms took billions of years to form and governed by the laws of physical interactions that are undetectable by modern instruments like all dark matter and dark energy. And who said that there are no upper and lower membranes of earth electromagnetic dark plasma halo that we call heaven and hell for the souls? After all, over 5 billions of humans who currently live on earth and over 50 billions who lived previously believed in forms of heaven and hell, therefore, this knowledge is imprinted deeply inside earth global information field, making heaven and hell real and actual thing inside earth electromagnetic halo layers.

Those dark matter plasma life forms inhabiting our bodies can be both creative and destructive but living in symbiosis with our brains, our actions seems to feed those plasma souls until they separated from the bodies upon brain death. Apparently, saints and divine entities had those electromagnetic plasma bodies so strong that they were actually glowing interacting with other electric particles in the air. This bodily glows were depicted in many church icons and frescos. The same phenomena was observed today in several "indigo children."

Thanks to the advancement in neuralgia, it was recorded several times in different hospitals brain encephalographs: when patient heart stopped and brain functions stopped there was the last and powerful spike of brain electricity and as well as faint undetectable to a naked eye electromagnetic glow. This is the exact definition of the soul leaving the body according to all religions. Now we have a recorded scientific proof of soul and the fact that our body is just a vessel for a higher electromagnetic plasma entity.

In other words, all biological bodies have plasma life forms, but not all plasma life forms have biological bodies. Furthermore, life-forms which have been described as "ghosts", "angels" and "deities" and other similar entities as weird plasma-based life-forms which are also evolving within Earth's dark matter field as well as other stars and planets dark matter fields. This field is what Star Wars would call "The Force" and religions would call "God" and atheists would call "Life force."

Doctor Jay Alfred: "We will realize that most of the encounters with aliens or ghosts are really sporadic encounters with beings from parallel interpenetrating Earth-based magma spheres. These aliens are dark matter entities...which generally do not register on the known electromagnetic spectrum - but can do so if their frequencies are brought down..."

Long lived or nearly eternal complex dark plasma life forms move into the biological entities generally by neutralizing or reversing their electric charge. More primitive short lived are redistributing the electric charges relative to the background electric field - high voltage wires, lighting strikes, heavy machinery, forming lighting balls and other unexplained atmospheric and space phenomena. When they fall in energy sporadically and are visible to us or our scientific instruments in the lower atmosphere, fast-moving plasma life forms will generate thunder balls.

The Earth's magnetosphere is composed almost entirely of ionized plasma and is dominated by the Earth's magnetic field. Fast moving plasma life forms would, therefore, be expected to generate a lot of interactions waves in the magnetosphere and sometimes be visible in the atmosphere.

Near-death experiences, astral traveling, alien encounters, angelic visitations, apparitions of saints or deities - both privately to individuals and to the public mass sightings occur in this higher energy dark matter of earth magnetosphere by those Earth information field plasma electromagnetic entities.

There is one important thing that you need to understand: all those electromagnetic plasma entities are not independent things - they all are part of Earth dark plasma halo and earth halo is part of Universal global information field, while Universe is most certainly part of something even bigger. Plasma life is the dominant life in the universe - not biological life. Plasma is not just a fourth state of matter it an intermediary state between pure matter and pure energy. It shares many characteristics of both. The plasma constantly regenerates itself and enclosed impurities in the same way that a living organism, like the unicellular amoeba, might encase a foreign substance. Plasma possesses traits of living things. When the carbon-based bodies die, symbiotic bio-plasma bodies continue their existence.

## Chapter 10. Islam and Judeo-Christian Chronicles.

The famous Islamic cosmographer and Persian physician who lived in the thirteenth century, Zakariya ibn Muhammad, states that jinns "are aerial animals, with transparent bodies which can assume various forms." This is recognized as one category of jinns which permeate the atmosphere and is reminiscent of Trevor Constable's plasma-based "sky creatures" and resembles our electromagnetic plasma life forms.

Gordon Creighton and Chris Line in 1989 argued that UFOs are in reality jinns in different issues of the "Flying Saucer Review". The idea that Earth's atmosphere could be the habitat of living aerial creatures which manifested as UFOs is not a new one. Charles Fort seemed to believe that too and Kenneth Arnold, who kick-started modern UFOlogy in 1947, also believed UFOs were living creatures. His belief that UFOs were space animals with the ability to change their density and shape-shift, has bothered other Ufologists.

Muslims believe that jinns have the power to fly and shape-shift by fitting into any space. It is interesting to note that the popular depictions of a "genie" often show a large giant whose body tapers into a vortex and who, despite his size, is able to squeeze into Aladdin's lamp or a small bottle. This piece of fiction actually illustrates quite nicely the ability of jinns to shape-shift. The depiction of a flame in a lamp is also reminiscent of plasma. Neon signs and fluorescent lamps have plasmas in them. Any type of fire with a very high temperature may generate plasma.

The Dictionary of Islam by Thomas Patrick Hughes states: "They become invisible at please (by a rapid extension or rarefaction of the particles which compose them), or suddenly disappear in the earth or air, or through a solid wall." Particles in plasma, through magnetic and electric forces, can increase their inter-particle distance to decrease the density of the plasma. All of these characteristics are consistent with electromagnetic plasma life forms. Hundreds of years ago the concept of plasma did not exist. But the term "smokeless fire" captures rather nicely the image of a plasma. If we had fluorescent lamps and neon signs composed of plasma a thousand years ago, they would probably be described as "smokeless fire" or "fire without smoke".

In Judeo-Christian chronicles "Moses came nearer to Mount Horeb he looked up to see a strange sight - in the distance a bush seemed to be on fire. Yet the bush was not consumed by the fire - it was a "smokeless fire". The Bible states "And the angel of the LORD appeared to him in a flame of fire out of the midst of a bush; and he looked, and lo, the bush was burning, yet it was not consumed." As Moses approached, God tells Moses to remove his sandals as the ground he was standing on was holy." The "burning bush" has all the hallmarks of cold plasma - like the auroras which appear around Earth poles. Important fact that Moses was asked to take off his sandals because he was told he was standing on the holy ground. Or was this because his sandals acted as an insulator between the ground and his body? The Bible also tells us that "his appearance changed after he came down from the mountain. He acquired a glow after the encounter" because his body was "charged" and the air around him was ionized or because one on new dark plasma electromagnetic entity entered him?

During the miracle at Fatima, Portugal in 1917 recognized by Catholic church there were many (Christians and Non-Christians) who witnessed glowing objects that were cross-shaped in the sky. Some even saw "white cups" which were described as "doves". Small white bodies which were "white as snow" were also interpreted as "doves". These "doves" resemble plasma formations and have also been seen in other Marian apparitions. These "doves" have been documented and photographed.

Over the Coptic Church of St Mary at Zeitoun - on the domes of the church for up to two hours or more at a time, always at night, a lady appeared in glittering light - so bright that it streamed across the church. She was often preceded or accompanied by luminous "doves", "strange bird-like creatures made of light", which did not flap their wings but glided. The figure moved across the domes bowing and greeting the enormous crowds, estimated at times to be as many as 250,000 people. The "doves" were consistently mentioned by eyewitnesses.

"Doves" also appear in the New Testament of the Bible - including the "dove" that appeared during the baptism of Jesus by John the Baptist in the Jordan River. The Bible states, "Now when all the people were baptized, and when Jesus also had been baptized and was praying, the heaven was opened and the Holy Spirit descended upon him in bodily form, as a dove..."

In addition to Famous Marian - Virgin Mary apparitional events of Zeitoun, Egypt and Fatima, Portugal where thousands observed the apparitions, and in the case of the Zeitoun appearances "intelligent forms" were recorded and authenticated photographically, there are countless prophecies told to those who had a special and unique relationship with these life forms have come to pass.

Bible and Quran did not record lies, they described electromagnetic plasma life forms that were present on earth long before humans and will be present long after us.

## Chapter 11. UFO & Aliens.

UFOlogist and even some respectable scientists tell us: there are 3 alleged types of aliens visitation: by so-called Gray aliens in saucer shaped crafts, Reptilians or Reptiloids in Sci-Fi style and cigar shaped ships and energy balls. Lots and lots of energy balls - the most common alien signings are those plasma balls.

Existence of Grays is proven by many witnesses including high ranking military officers. However, there is one big problem with Grays - they look too similar to humans: one head, two eyes, four limbs, even closely matching number of fingers. It's simply impossible that intelligent life at the distant alien star would be so similar to humans. It means that Grays are not aliens at all. Grays traveled from the future to observe their evolution and collect the historical evidence. In other words, gray aliens are future humans and they do travel though time but no so much though galactic distances. They saucers shaped UFO are actually time and space bending time machines.

DARPA scientists had alien UFO engine in their possession for at least 50 years stored in Sandia Lab in New Mexico. Finally, they were able to explain how it works and even make a crude copy. What I tell you is very highly classified. First of all, I will try to explain as simple as I can since full understanding requires studying of post Doctorate physics. Second of all, it's like nowadays Columbus would travel to Indians on nuclear submarine and Indians would try to copy it.

The most important thing to remember - the whole engine is powered Liquid Metallic Hydrogen. To be liquid at room temperature metallic hydrogen has to be under 4,000,000 atm pressure. Only then hydrogen atoms would display metallic properties, losing hold over their electrons. Such pressure is not possible to achieve even for alien technology so they used Pulsed lasers superheating to get hydrogen in metallic state under only 250,000 atm. Scientists from secret DARPA branch at Z Pulsed Power Facility (known as Z machine) build this device in 2015 in Sandia's lab in New Mexico copying alien engine and since than it was the best kept government secret. Officially, they said they work on atomic metallic hydrogen that could be used as the most powerful chemical rocket fuel, as the atoms recombine to form molecular hydrogen. Which is true. It has 5 times more exhaust velocity then any kind of Chemical rocket that could outperform Nuclear rocket as well. But it gets stranger. Superconductor metallic hydrogen fluid also is in quantum state. Quantum state fluid theoretically isolate the whole system from its surroundings and known to generate super currents within it. Such fluid has highly unusual reactions to external magnetic fields and rotations. In other words, it can bent space and time around like a miniature black hole allowing any spaceship powered by such device if not travel trough time like a time machine but at least bent space and time around it shortening the travel distance in thousands of times and enabling it appear of nowhere like many UFO observers reported.

Therefore, we are dealing with time traveling future humans - not aliens. While Reptilians - if they really exist, because unlike Grays they leave nearly 0% proof of their visitations, except some historical records of humans worshipping serpent gods in the ancient past - are indeed very different species. Assuming they exist, some say they came from Orion constellation. It might be the cause but then they should have achieved an unimaginable level of technology advancement including travel way faster that the speed of light - which is physically impossible (even theoretically). It's much more likely. however, that they also come from earth, but very different earth - were asteroid did not trigger mass dinosaur extinction and in millions of years they developed in much more advanced space and time travel species.

There is a beautiful legend why Reptilians are coming to Earth and it has to do with the yellow metal:

Millennia fly by upon the Earth. Hundreds of generations changed. So many things changed. Yet, one thing remains unchangeable. Mankind love for gold. Indeed, why are we fascinated with this yellow metal? But the right question to ask would be why aliens made us love it. Do you know that bacteria exists that produces 24K or 99.9% pure gold. Yes, that bacteria literally poops the gold nuggets by turning gold chloride—a toxic chemical liquid you abundant in nature—into 99.9% solid gold. In fact, Michigan state university has an entire lab filled with gold producing bacteria. I bet you didn't knew about it so read more and you will find facts about gold that including why aliens use it and where is city of gold located. For that bacteria gold is toxic waste, because gold kills bacteria but their poop is our treasure. In fact, if your body is compressed each of us contain 0.2 milligrams of gold.

Most of the gold is located in heart. So it's true that some of us have hearts of gold. And if that gold is extracted from all humans - it will be more gold than in the entire Unites States gold reserve. Gold, unlike bitcoin, is real physical thing. It is also relatively easy to mold producing the shapes that you want. You can use it for electronic conductors, you can wear it, you can eat it, since it has healing effect on human body by gold nano particles eliminating harmful bacteria. Several restaurants around the world serve you gold to eat. But each dish costs hundreds of dollars. I bet you are getting hungry just thinking about it. See how easy it's to manipulate human mind into something. It was even easier for aliens to manipulate early humans to love and adore gold.

Archeologist found an ancient perfect 24K carat gold disk. It first appeared two and half thousands years ago in Rome before Roman Republic or Empire even existed. It is so perfect that it would be hard to produce even using the modern technology. In fact, it was made by Anunnaki aliens hundreds of thousands years ago along with many other gold artifacts. Alien contacts started millions of years ago when first Anunnaki aliens came to mine gold and other metals from earth.

Why did they came all the way here? Because unlike most of other small planets Earth suffered a catastrophic collision with planet Thea in the past that nearly destroyed Earth and created Moon but it also throw gold and other heavy metals on earth surface from it's core, making Earth so desirable for mining when on other planets aliens had to drill the core to get molten heavy metals, while on earth they were cool and ready right under their feet or tentacles. To assist in gold mining and service Anunnaki played with DNA of great apes to create what he know as humans today. So the point to Charles Darwin here - we are great apes descendants. Once they minded most of it - they left - leaving a few pyramids constructed by humans as a reminder of ancient aliens landmarks.    And they also left one thing on Earth behind - our genetic love for gold. That's why to this day we are fascinated with gold and literally worshipping gold. Most of Earth deities are made from gold - from gold crosses on top of the churches to the gold Buddha. And you don't even suspect that those gold idols are actually representing aliens that visited earth and genetically modified us to love gold in order to find it and mine it for their alien needs. Only a few selected humans were chosen to wear gold throughout our history and they also were worshipped like Gods.

Aliens didn't need much gold for their space ships because as building material gold is impractical because its very heavy and very soft in its pure form. Spaceship made of gold would not fly far. It's only humans who can build their cars from gold. Even this small car made of gold weight several tons. Aliens needed gold to use on their skins and in their atmosphere, since it's the only antibacterial substance that does not have any side effect and can be used on humans, animals or aliens bodies. You can breath it, eat it, wear it and, unlike other metals, it will only make you healthier.

Aliens used gold nano particles in their atmosphere because it was blocking harmful Ultraviolet rays of their star, deflecting star light to prevent global warming of their planet, making their air healthier to breath, and killed all harmful bacteria in the air and on the surface. Now you understand that gold has universal value but we humans using it the mostly wrong way by hoarding it in bars and coins instead of using it for our health needs and to prevent global warming by making gold nano particles deflecting part of sunlight and harmful UV rays.

This is a very compelling story, but there is no 100% solid proof to it. In the best case scenario, we just have two time travelling species visiting earth for the several last millennia: Gray and Reptilians and virtually zero intergalactic aliens.

Only about 100 people ever saw Gray alien on military base, and virtually no one can prove that he or she saw a Reptilian. What is proven that countless millions of people saw and can prove though videos and collaborating statements of people who never saw each other - are plasma orbs and alien balls. Those aliens that part of Earth and Universal electromagnetic information field. They would be far more adaptable then our current carbon based life forms as we know it. Plasma is by far the most prevalent form of matter in the universe, and it exists everywhere. Since it's not impossible that plasma life can form naturally, atmospheric life forms might exist which are based on aerogels and electrostatics rather than on water-gels and ionic chemistry. The gas based analog to ionic chemistry in solutions is plasma chemistry.

The existence of plasma organisms would explain some reports of distant moving lights at night, encounters with "will-o-the-wisps," bioluminescent patterns in the air over the Indian ocean, and

41

reports of ball lightning when thunderstorms are lacking. Plasma creatures would be as invisible as an underwater ice-cube, and might only be detectable when it causes slight optical distortion of background objects.

Indeed, most of what we read and see describe the UFOs as glowing orbs. There are many NASA videos that show these glowing orbs in space, moving in a shimmer that suggests that they are going where they want to go, that the movement is controlled. Most of the reports you here from pilots, especially all the foo fighter reports, which are proven to be real, describe glowing orbs. This is because most of these reports are of some sort of plasma life forms.

## Chapter 12. Electromagnetic Brain.

This book became possible largely because of the works of David Bohm in 1940s.

Doctor David Bohm worked at the Lawrence Radiation Laboratory where, after receiving his doctorate in 1943, he began what was to become his landmark work on plasmas. Bohm was surprised to find that once electrons were in a plasma, they stopped behaving like individuals and started behaving as if they were part of a larger and interconnected whole. He later remarked that he frequently had the impression that the sea of electrons was alive.

In 1951 Bohm wrote a classic textbook titled Quantum Theory, in which he presented a clear account of the orthodox, Copenhagen interpretation of quantum physics. The Copenhagen interpretation was formulated mainly by Niels Bohr and Werner Heisenberg in the 1920s and is still highly influential today.

Bohm sent copies of his textbook to Bohr and Einstein. Bohr did not respond, but Einstein phoned him to say that he wanted to discuss it with him. Later, Einstein enthusiastically told Bohm that he had never seen quantum theory presented so clearly, and admitted that he was just as dissatisfied with the orthodox approach as Bohm was. They both admired quantum theory's ability to predict phenomena, but could not accept that it was complete and that it was impossible to arrive at any clearer understanding of what was going on in the quantum realm. Bohm was the father of plasma physics theory.

The plasma physics theory model explains a wide variety of phenomena associated with genuine sightings of aliens, ghosts, deities, angels by both individuals and groups - using an internally consistent scientific framework. These sightings show the presence of plasma-based life forms originating from global information flied. These plasma aliens such as ghosts, fairies, grays, deities or angels share common properties and even shed light on the evolution of carbon-based life and consciousness on the Earth that we are most familiar with.

During a direct reading of the Earth information field known as Akashic records a current is generated in the brain of the reader's bioplasma body which connects as a cord to Earth's brain allowing information to be transferred. This is similar to the "silver cord", frequently reported by Out-of-Body experiencers, that connects the physical body to the bioplasma body. In a sense, this is like connecting to Wi-Fi internet, except that we are connecting to a much larger and older information field with one big exception: Different computers connect to Wi-Fi generating the same virtual world while different brains generate different worlds. This means that each person who accesses Earth's brain would contextualize the information received to his or her brain in terms of his or her own understanding and expectations. However, it also means that Earth's brain does not process inputs in exactly the same way as human brains. Therefore, an individual's memories may be modulated by Earth's brain. Furthermore, Earth's brain constructs space and time differently. This supports the claim by some that the information received in dreams and premonitions is ultimately derived from Earth's global information field. Each living brain is a functioning link or nucleus in Earth's electromagnetic plasma brain.

If Earth has an electromagnetic brain, where does the planet obtain its sensory inputs? Earth does it by generating life-forms. The myriad of life-forms (including humans) on Earth are in fact the many eyes and ears of that "brain". It is in the interest of global information field to generate life-forms so that it can see, hear, taste, touch, smell, sense and become aware of itself and developed its own memories. These include the memories of every person deeds who has ever lived on this planet. Therefore, your life's memories are not solely yours – it is also Earth's and the rest of humanity. It also helps the brains that protect life, defend animals and other life forms by recycling their dark plasma into new brains and new bodies causing the phenomena of reincarnation and annihilate the components whose mission was to decrease life and case harm to the planet and it's species, filling religions concept of hell and heaven.

If you enjoyed this Igor Kryan book, consider the same author books titled "Earth before the Pyramids" (2009) Creator's Riddle: Darwin vs. God (2012) and Apocalypse of Magdalene and Judas: Everything Church Does Not Want You To Know (2016).

**Sources.**

"Could Life be based on Silicon rather than Carbon?" NASA Astrobiology Institute.
http://nai.nasa.gov/astrobio/feat_questions/silicon_life.cfm

"Physicists Discover Inorganic Dust With Lifelike Qualities." Science Daily. Aug. 15, 2007.
http://www.sciencedaily.com/releases/2007/08/070814150630.htm

"Plasmas - the Fourth State of Matter." Princeton Plasma Physics Laboratory. Jan. 18, 2001.
http://fusedweb.pppl.gov/CPEP/Chart_Pages/5.Plasma4StateMatter.html

Battersby, Stephen. "Could alien life exist in the form of DNA-shaped dust?" NewScientist.com. Aug. 10, 2007.
http://space.newscientist.com/article/dn12466

Battersby, Stephen. "Invasion of the Plasmozoids." New Scientist Space Blog. Aug. 10, 2007.
http://www.newscientist.com/blog/space/2007/08/invasion-of-plasmozoids.html

Beradelli, Phil. "From Space Dust to Spacefarers." ScienceNow Daily News. Aug. 14, 2007.
http://sciencenow.sciencemag.org/cgi/content/full/2007/814/2

Booth, Robert. "Dust 'comes alive' in space." The Sunday Times. Aug. 12, 2007.
http://www.timesonline.co.uk/tol/news/uk/article2241753.ece

Mullen, Leslie. "Plasma, Plasma, Everywhere." Science@NASA.
http://science.nasa.gov/NEWHOME/headlines/ast07sep99_1.htm

Than, Ker. "Hot gas in space mimics life." Space.com. USA Today. Aug. 14, 2007. http://www.usatoday.com/tech/science/space/2007-08-14-hot-gas-like-life_N.htm

Tsytovich, V.N., Morfill, G.E., Fortov, V.E., Gusein-Zade, N.G., Klumov, B.A. and Vladimirov, S.V. "From plasma crystals and helical structures towards inorganic iving matter." New Journal of Physics. Aug. 14, 2007. http://www.iop.org/EJ/abstract/1367-2630/9/8/263

Zimmer, Carl. "Scientists Urge a Search for Life Not as We Know It." New York Times. July 7, 2007. http://www.nytimes.com/2007/07/07/science/space/07alien.html?ex=1187323200&en=36fc1468d6f9dc34&ei=5070

Penfield, W., Mystery of the Mind, Princeton University Press, 1975.

Hebb, D. O., Organization of Behavior, A Neuropsychological Theory, Lawrence Erlbaum, 2002. (First published in 1949.)

Uehara, M. et al, Physics and Biology: Bio-plasma physics, American Journal of Physics 68 450-455, 2000.

Russell, P., The Brain Book, Routledge & Kegan Paul PLC, 1979.

Radin, D., Entangled Minds, Paraview Pocket Books, 2006.

Richards, T.L., Kozak L., Johnson L.C., Standish L.J., Replicable functional
magnetic resonance imaging evidence of correlated brain signals between
physically and sensory isolated subjects. Journal of Alternative and Complementary Medicine 11, 955-63, 2005.

Charman, R.A. (2006). Has direct brain to brain communication been demonstrated? Journal of the Society for Psychical Research, 70, 1-24.

Charman, R.A. (2006). Direct brain to brain communication - further evidence from EEG and fMRI studies. Paranormal Review, 40, 3-9.

Charman, R.A. (2006). Correspondence: Something really is going on. Journal of the Society for Psychical Research, 70, 249-251.

Morse, M. and Perry, P., Where God Lives, The Science of the

Paranormal and How our Brains are Linked to the Universe, HarperSan Francisco, 2000.

Yogananda, P., Autobiography of a Yogi, Self-Realization Fellowship, 1946.

Alfred, J., Aliens from Dark Earth, New Dawn Magazine, Australia, May-June 2009.

Alfred, J., Earth's Brain, Akashic Records and Paranormal Imprints, May 2007.

Alfred, J., Brains and Realities, Trafford Publishing, Canada, 2006.

## About the Author.

Igor Kryan was born in 1979 in Kiev, Ukraine. He arrived to the Unites States in 1999 and his daughter Alisa was born in 2008. The author graduated with Honors at the National Ukrainian Medical University (AA Degree in Health Science - 1999), San Francisco State University (BS degree in Biology - 2003), New York Vernell University (Master of Fine Arts - 2003), Amsterdam Ravenhurst University, Netherlands, (MS degree in Cell and Molecular Science - 2004), Canterbury University, UK and the Ukrainian National University (PhD Doctorate degree in Astrobiology - 2010).

Igor Kryan is famous in the United States and Europe for over 1,000 original artworks and about a dozen of published English books such as "The Source" in 2004, "Future History" in 2006, "History of the Impossible" in 2008, "Earth before the Pyramids" in 2009, "2012: Hoax or Shock?" in 2010, "What if the British had Won" in 2011, "Angels and Demons Art Trilogy" in 2012, "Creator's Riddle: Darwin vs. God" in 2012, "Apocalypse of Magdalene and Judas" in 2016 and

"The Second Coming: Jesus Arrived but Goverment Hid Him" and Alien Life and Dark Plasma:
What Makes You Alive and Self Aware?
in 2018 as well as several satirical books in Russian. Most of Dr. Kryan's books enjoy overwhelming success - hundreds of thousands electronic and digital copies along with thousands of actual paper copies were sold and many more are being currently sold across the globe.

Doctor Kryan is known for: Human and Animal Liberation Activism. Numerous books and publications. Professional paintings and drawings. Research of solar activity and UV radiation effects. Research of ancient and modern history and anthropology in conjunction with future analysis including prediction of 2008 Great Recession and New Cold War as early as 2005, Origin of life theory (2003), 50 consecutive miracles theory (2009), Universal Simulation Creator Theory (2012).

However, Doctor Kryan was not known for his association with CIA, FSB and ALF until he decided to go public in 2014 in order to save America from CIA sponsored modern day sophisticated enslavement and establishment of totalitarian state. The newest book CIA Trilogy: CIA Millennium Hilton (2013), CIA Earth Blood (2015), and CIA Oblivion (est. 2020) reveals both unknown author biography serving CIA, FSB and ALF interests and devious CIA plan to replace free world we love and cherish with totalitarian super state making us all obedient slaves in the process.

IK 2015